Photobiological Studies No.2

PLANT HISTOLOGY

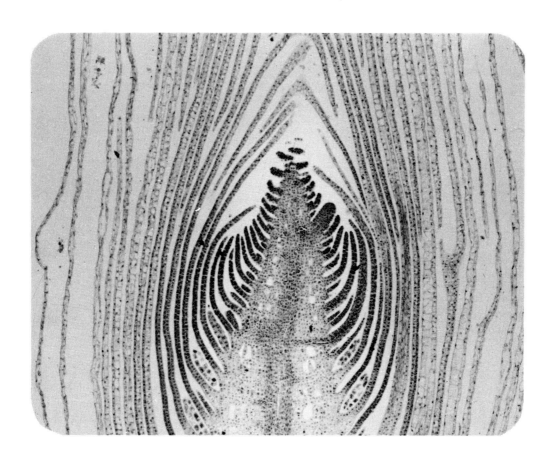

K. G. BROCKLEHURST M.A., M.I.Biol.
Senior Biologist, Warwick School

B. JOAN DAVIES B.Sc.
Warwick School

HODDER AND STOUGHTON
LONDON   SYDNEY   AUCKLAND   TORONTO

Photobiological Studies

Making observations on which to build a basic knowledge of biology has always been a problem for students. Teachers have approached this problem using a wide variety of equipment and methods and, in the last decade, with a considerable degree of success. This series of photo-biological studies is offered as a further contribution toward solving the problem in areas where the material is difficult to obtain or observe in detail. It is based on the belief that good quality photographic material is valuable for the students as a corrective to the severe simplification of textbook diagrams and as an aid to practical work and rapid revision. The text is minimal, being confined to essential explanatory information and questions of a leading nature. We are including test papers in the series because we have found photographic data to be valuable in examinations, especially where practical experience is under test.

Cover photomicrograph: T.S. Stele and part of cortex of Iris root.

Title page photomicrograph: L.S. stem apex of Canadian pondweed (*Elodea canadensis*).

ISBN 0 340 22051 1
Copyright © 1978 K. G. Brocklehurst & B. Joan Davies.

Printed in Great Britain for Hodder and Stoughton Educational, a division of Hodder and Stoughton Ltd., Mill Road, Dunton Green, Sevenoaks, Kent, by Morrison & Gibb Ltd., London and Edinburgh.

Contents

INTRODUCTION	2
CELLS	3
MORE CELLS	4
SPECIALISED CELLS—SIEVE-TUBES AND VESSELS	5
YOUNG ROOT OF DICOTYLEDON—TISSUE DISTRIBUTION	6
CELL DETAIL OF STELE	7
YOUNG ROOT OF MONOCOTYLEDON	8
CONDUCTING TISSUES	9
YOUNG STEM OF DICOTYLEDON	10, 11
YOUNG SUNFLOWER STEM	12
STEM OF VEGETABLE MARROW	13
CONDUCTING TISSUES	14
STEM OF MONOCOTYLEDON	15
LEAF OF DICOTYLEDON	16
LEAF TISSUES	17
STOMATA	18
STOMATA AND MESOPHYLL	19
WOODY STEM, T.S.	20
WOODY STEM, L.S.	21
ELDER STEM	22
BARK AND LENTICELS	23
SPECIALISED LEAVES	24

Introduction

This collection of photomicrographs has been selected to supplement and assist practical classes in the vegetative anatomy of flowering plants and is intended for students working at an elementary level. A further book is planned to illustrate the reproductive anatomy of flowering plants and to include mitosis and meiosis. It is hoped that the questions, which are numbered in sequence on each pair of facing pages, will encourage students to look at microscopic detail carefully, to interpret what they see and to try to relate it to function. None of the material used is difficult to obtain, but occasionally we have had to select from a number of similar preparations in order to illustrate particular points clearly. Most of the photomicrographs are of stained sectioned material using bright-field microscopy; wherever phase-contrast or polarised light was employed this is indicated after the caption by [Ph-c] or [Pol] respectively. All the photographic material has been specially made for this book.

B.J.D. & K.G.B.

Cells

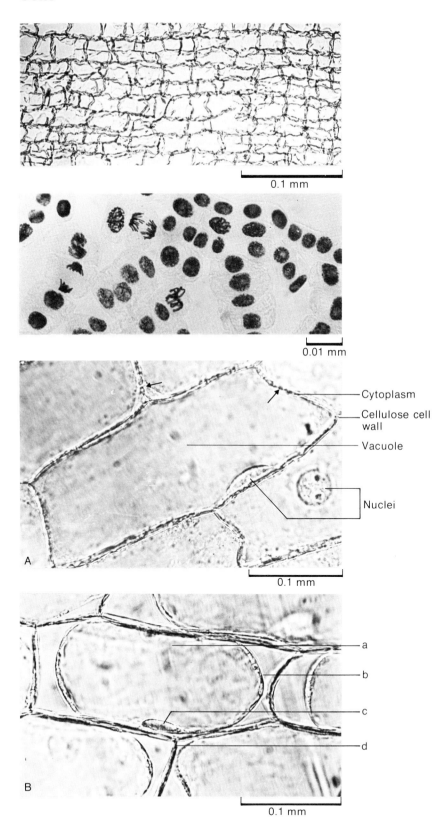

Figure 1 A thin slice, or section, of cork (the bark of the cork oak, *Quercus suber*). Robert Hooke, in 1663, was the first to see these structural units called cells.

1 How many layers of cells are in the section?
2 Are all the cell walls intact?

Figure 2 A piece of root tip of broad bean (*Vicia faba*) squashed to spread the cells and stained in aceto-orcein. Nuclei appear dark, cytoplasm is pale and unstained and cell walls are difficult to see since they too are unstained.

3 By comparison are the cork cells of Fig. 1 empty?

Figure 3A Living epidermal cells peeled from the inside of an onion-bulb scale and mounted in water. Read the labels carefully. If you cut down the light intensity on your microscope you will probably get this 'contrasty' effect.

4 What do living cells have which dead cells do not?
5 To what do the two arrows point?

Figure 3B Similar onion cells immersed in 10% glucose solution. Identify parts a to d.

6 What part of the cell has shrunk and why?
7 Can you now see cytoplasm and vacuole more clearly?
8 Where is the cytoplasmic membrane?
9 Which cells, A or B, are turgid and which are plasmolysed?
10 How does a turgid tissue differ mechanically from one which is plasmolysed?

3

More Cells

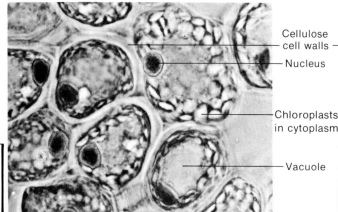

Cellulose cell walls

Nucleus

Chloroplasts in cytoplasm

Vacuole

Figure 4   Chlorenchymatous cells from the cortex of a stained T.S. of the stem of mistletoe (*Viscum album*). [Ph-c]
1   What is the value of a green stem to a plant?
2   Which parts of the cell contain chlorophyll?

50 μm

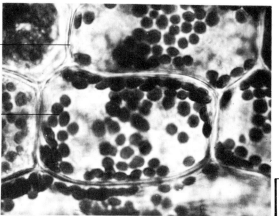

Cellulose cell walls

Chloroplasts in cytoplasm

Vacuole

10 μm

Figure 5   A whole live cell from the leaf of Canadian pondweed (*Elodea canadensis*).
3   Can you detect a vacuole in the cell?
4   Where in the cell was the microscope focused?

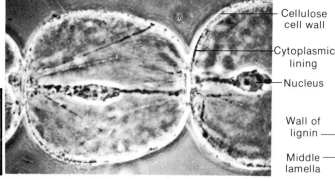

Cellulose cell wall

Cytoplasmic lining

Nucleus

50 μm

Figure 6   A whole live cell from staminal hair of *Tradescantia sp.* The cytoplasmic strands and lining exhibit streaming.   [Ph-c]

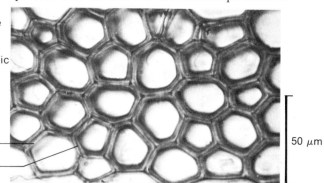

Wall of lignin

Middle lamella

50 μm

Figure 7A   T.S. stained fibres from sclerenchyma of the cortex of a marrow stem (see also Fig. 23A).
5   How do these cells differ from those in Fig. 4?

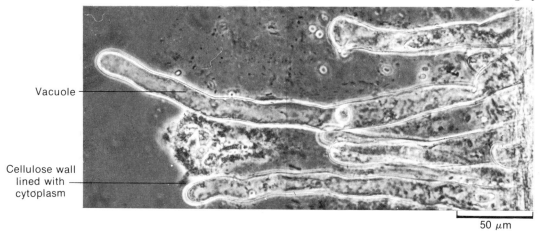

Vacuole

Cellulose wall lined with cytoplasm

50 μm

Figure 7B   A whole fibre from macerated oak wood.   [Pol]
6   Can you relate the whole cell to the sectional view?

Figure 8   A piece of young root of garden pea (*Pisum sativum*), mounted in water.   [Ph-c]
7   What is the function of root hairs?
8   How many root-hair cells have a clear nucleus?

4

Specialised Cells — Sieve-Tubes and Vessels

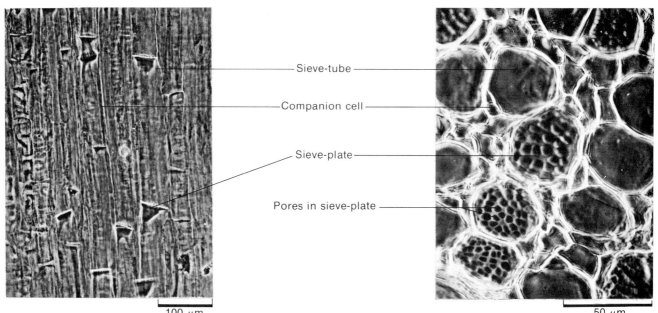

Figure 9   Sieve-tubes in a small part of a L.S. stem of marrow (*Cucurbita pepo*).   [Ph-c]

Figure 10   Sieve-tubes, some with sieve-plates, and companion cells in a small part of a T.S. marrow stem. [Ph-c]

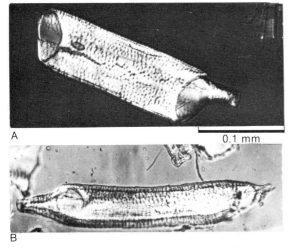

Figure 11A & B   Segments of vessels from macerated wood of oak. Note the open ends and the pits in the walls.   [Pol]

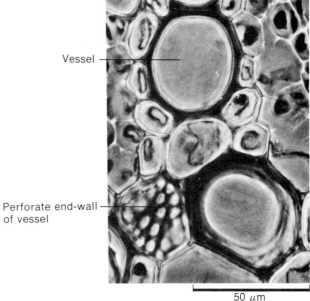

Figure 12   Three vessels in T.S. from a root of buttercup (*Ranunculus repens*). Note their thick lignified walls and in one case the perforate end wall lying obliquely to the plane of the section.   [Ph-c]

Both sieve-tubes from phloem and vessels from xylem (wood) are specialised for conduction and are, therefore, important parts of vascular tissues.

9   What substances are transported in each, and from where to where in the plant?

10   Each vessel segment forms from one long cell; what must happen to many end walls as a vessel develops?

11   Is a vessel, strictly speaking, a single cell?

12   How would you make sure you did not confuse a sieve-plate with the end wall of a vessel?

5

Young Root of Dicotyledon — Tissue Distribution

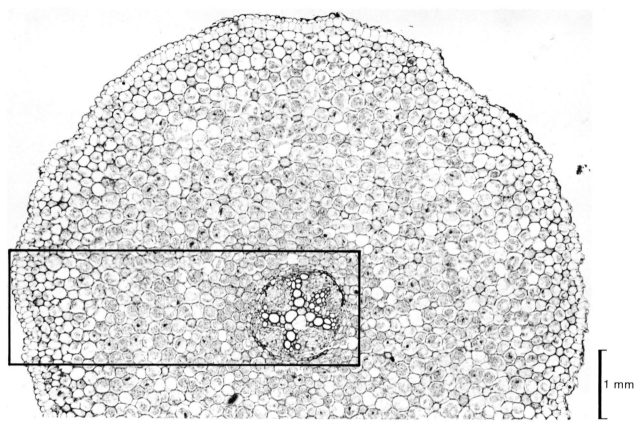

Figure 13A Young root of creeping buttercup (*Ranunculus repens*) sectioned transversely and stained.

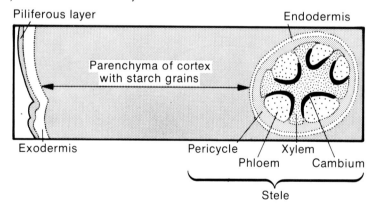

Figure 13B Tissue distribution diagram for a representative part of Fig. 13A. Note that stele, cortex and pericycle are names of regions, but the remainder are tissue names.

1 Can you suggest a reason for the apparent emptiness of some of the parenchymatous cells?

Study Figs. 13A & B and 14 checking each labelled item and then attempt the questions below.

2 How many xylem arms are in the stele?

3 Cambium cells can undergo active cell division. Can you locate places where this has occurred?

4 What do the daughter cells of cambium cells develop into?

5 The endodermis is far from completely developed. Where is lignin thickening only on the radial walls?

6 Which of the root tissues has numerous intercellular air spaces and what purpose might these spaces serve?

7 What is the value to the plant of so many starch grains?

8 How is the piliferous layer adapted to its function? Can you suggest a reason for root hairs not being visible on the section?

Cell Detail of Stele

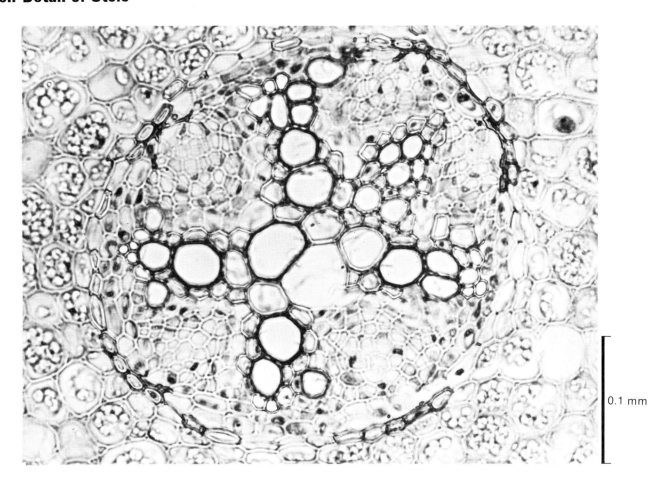

0.1 mm

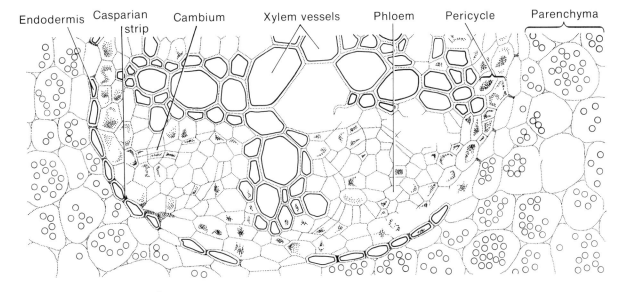

Endodermis Casparian strip Cambium Xylem vessels Phloem Pericycle Parenchyma

Figure 14   The central region of Fig. 13A more highly magnified.

Young Root of Monocotyledon

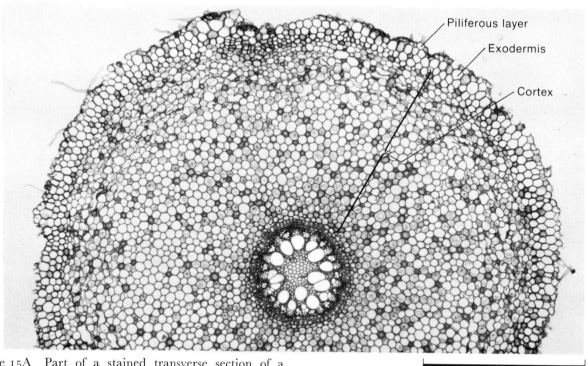

Figure 15A Part of a stained transverse section of a young root of *Iris sp.*

1 mm

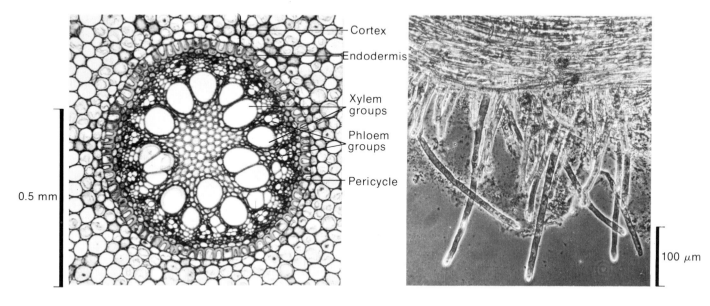

0.5 mm

100 μm

Figure 15B The stele of Fig. 15A more highly magnified. The light grey cell walls in B were red, having been stained by the dye safranin, which combines readily with lignin. The thinner cellulose cell walls were stained with aniline blue and appear blacker. Such double staining is often used to distinguish between cellulose and lignin.

Figure 16 Live root hairs from pea seedlings grown in paper rolls containing tap water. The root hairs have been laid flat and somewhat tangled in mounting. The speckled black debris among the hairs consists mainly of bacteria.   [Ph-c]

1 On what might the bacteria have been feeding?

Conducting Tissues

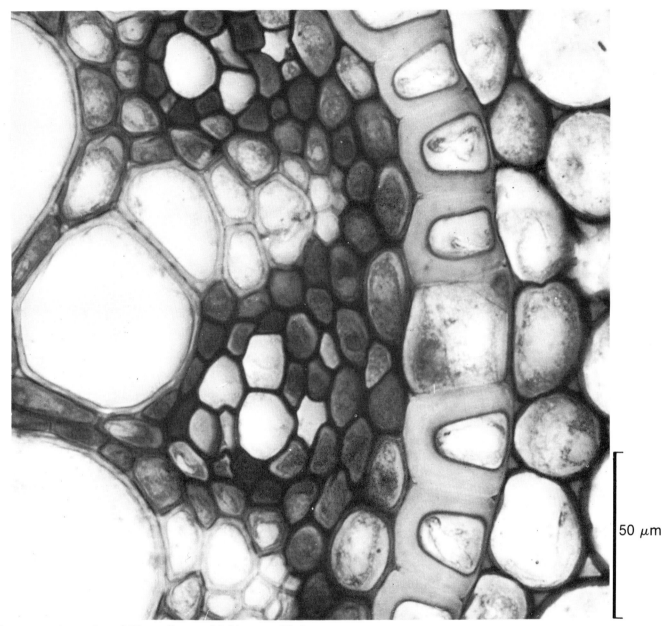

50 μm

Figure 17   A portion of Fig. 15B more highly magnified.
Determine which part of Fig. 15B is included.

The roots shown on pages 6 & 7 and 8 & 9 are typical of
dicotyledons and monocotyledons respectively.

2   Can you construct a table of differences between the
two noting details of piliferous layer, exodermis, cortex,
parenchyma, endodermis, number of phloem and
xylem groups, and distribution of xylem vessels? Use

Fig. 17 to help you and remember that sieve-tubes can
often appear empty in transverse section.

3   What is the three dimensional shape of the endodermis?
4   Can you suggest functions for the endodermis? Think
about (a) invasion by parasitic fungi from the soil and
(b) water transport.
5   What is the value of the six unlignified cells of the
endodermis—only one appears in Fig. 17.

Young Stem of Dicotyledon

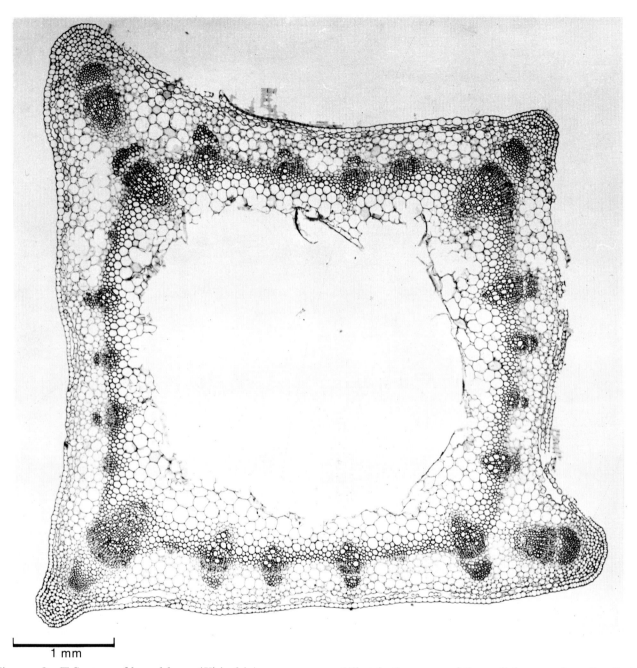

1 mm

Figure 18   T.S. stem of broad bean (*Vicia faba*).

1 Which corner of the stem and which vascular bundle are shown enlarged in Figs. 20 & 21?

2 After identifying all the tissues of the stem can you say in what ways this stem differs from the root in Fig. 13A?

3 Can you relate any of these differences to the different functions carried out by the root and the stem?

4 What is the cause of the radial rows of small cells in the vascular bundles (clearly seen in Fig. 21)?

5 What do you notice about the walls of the vessels in Fig. 21?

6 Collenchyma is a tissue appearing only in the corners of the stem—what, would you suggest, is its function? Compare Figs. 31A & B and 33.

7 Can you find evidence that broad bean stores starch as a food reserve (compare Fig. 14)?

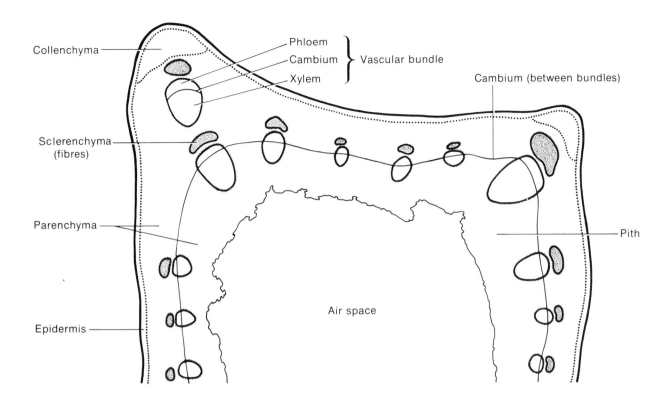

Figure 19   Tissue distribution diagram for Fig. 18.

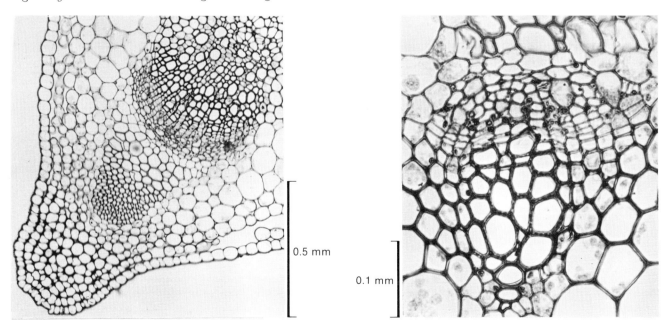

Figure 20   One corner of Fig. 18 enlarged.

Figure 21   One vascular bundle from Fig. 18 enlarged.

Young Sunflower Stem

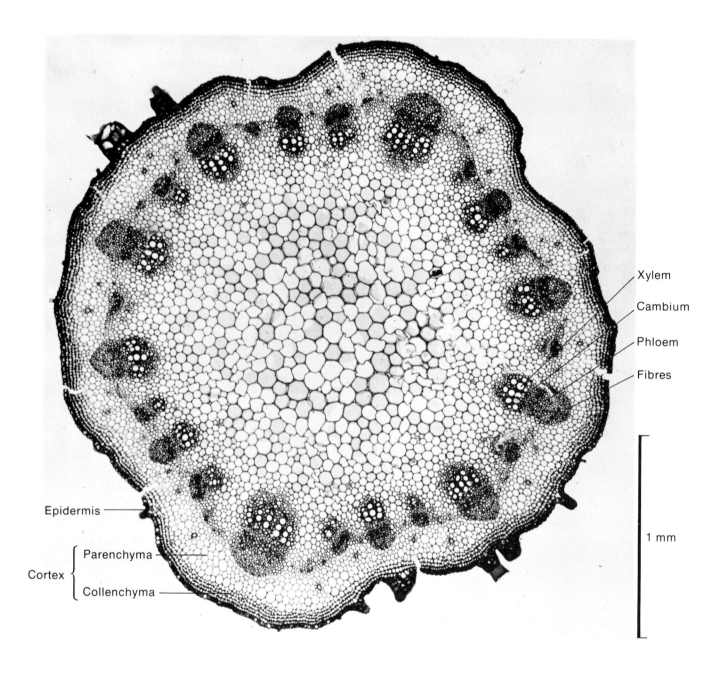

Labels on figure: Xylem, Cambium, Phloem, Fibres, Epidermis, Cortex { Parenchyma, Collenchyma }, 1 mm

Figure 22   Stained T.S. young stem of sunflower (*Helianthus annuus*).

1   Identify the vascular bundles of sunflower stem. Does their arrangement and type conform to that of broad bean (p. 10 and 11)?

2   What differences are there between the stem structures of sunflower and vegetable marrow (p. 13) as seen in transverse section? Can you relate these differences to their erect or trailing habits of growth?

3   With the help of the high-power detail on p. 14 identify the tissues cut in the longitudinal section in Fig. 23B and then work out the line on Fig. 23A which might have been included in the plane of the section.

Stem of Vegetable Marrow

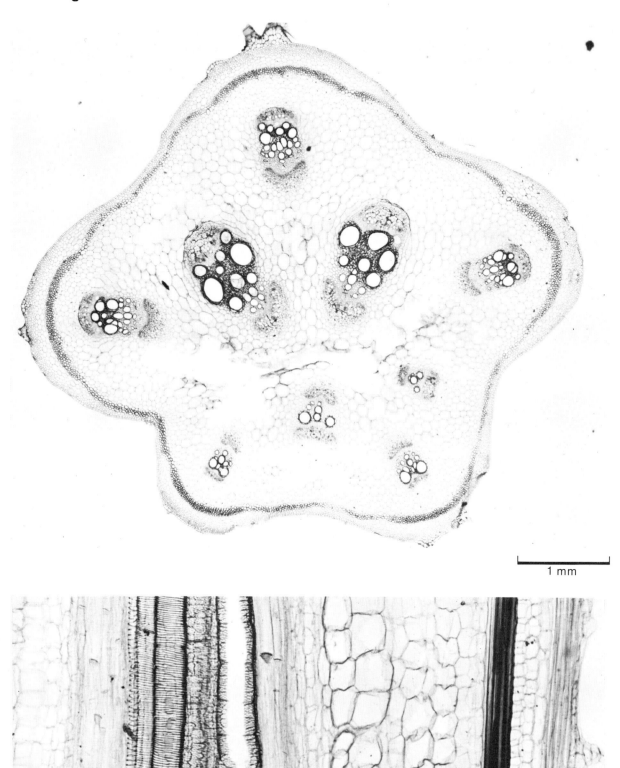

1 mm

Figure 23A & B   A—T.S. stem of marrow (*Cucurbita pepo*), stained. B—L.S. of part of the same.

Conducting Tissues

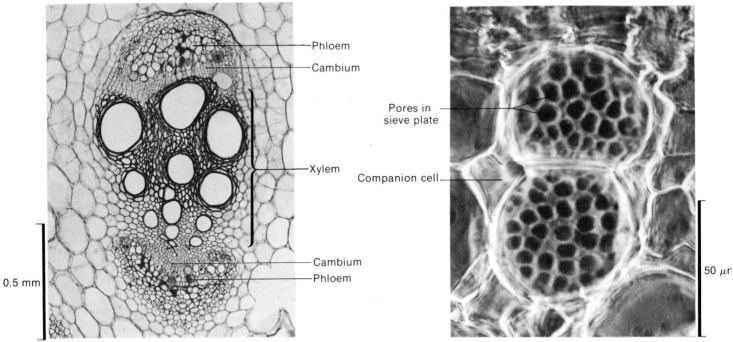

Figure 24   A vascular bundle of Fig. 23A more highly magnified.

Figure 25   Two sieve-plates from T.S. stem of marrow. [Ph-c]

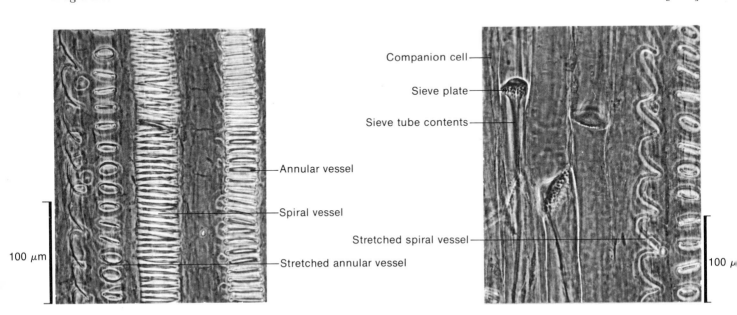

Figure 26   Spiral and annular vessels of marrow as seen in L.S. stem. [Ph-c]

Figure 27   Sieve-tubes and young stretched vessels in L.S. stem of marrow. [Ph-c]

1   How are sieve-tubes and vessels specialised for transport of water and nutrients?

Stem of Monocotyledon

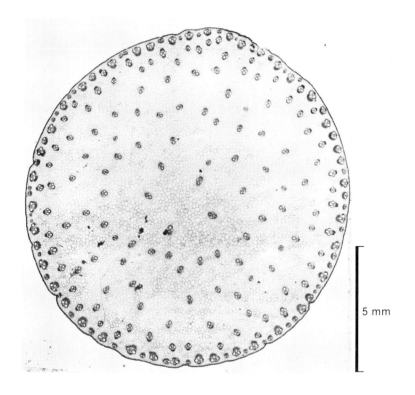

1 How are the vascular bundles arranged in a maize stem?
2 How many prominent vessels occur in each bundle?
3 Is the phloem near the inside or the outside of the vascular bundle?
4 The section in Fig. 30 is several cells thick. Bearing this in mind, can you work out the plane of the section relative to the transverse section of a bundle?
5 What gives the very tall maize stem strength to resist bending?

5 mm

Figure 28   Stained T.S. stem of maize (*Zea mais*).

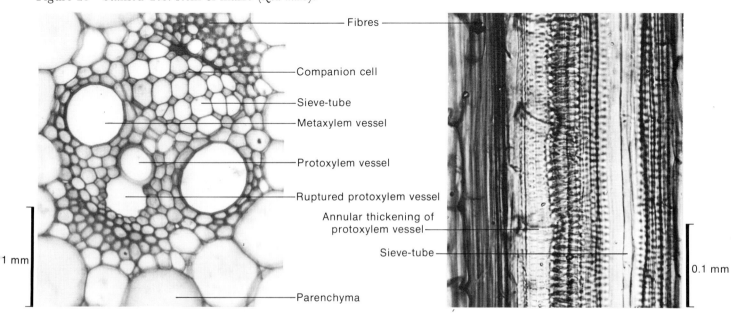

Fibres

Companion cell

Sieve-tube

Metaxylem vessel

Protoxylem vessel

Ruptured protoxylem vessel

Annular thickening of protoxylem vessel

Sieve-tube

1 mm

0.1 mm

Parenchyma

Figure 29   Stained T.S. vascular bundle of maize.   Figure 30   Stained L.S. vascular bundle of maize.

15

Leaf of Dicotyledon

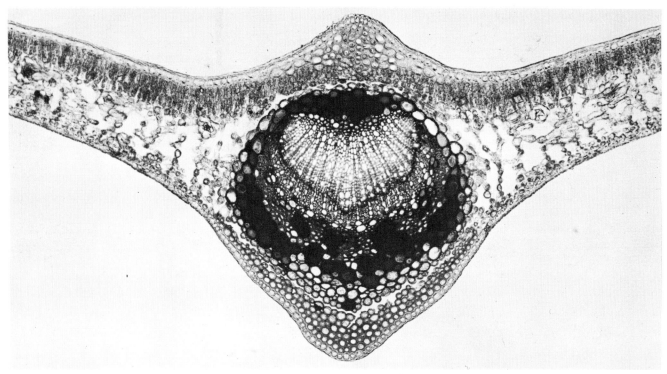

Figure 31A   T.S. leaf of cherry laurel (*Prunus laurocerasus*) passing through midrib and lamina.

1 mm

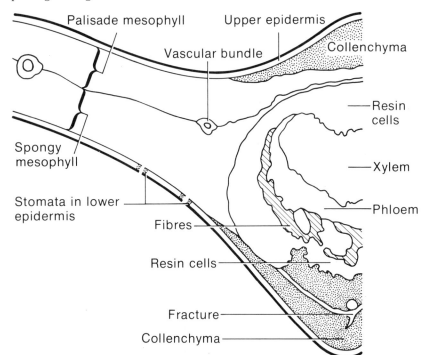

Palisade mesophyll

Upper epidermis

Vascular bundle

Collenchyma

Resin cells

Xylem

Phloem

Spongy mesophyll

Stomata in lower epidermis

Fibres

Resin cells

Fracture

Collenchyma

Figure 31B   Tissue distribution of Fig. 31A.

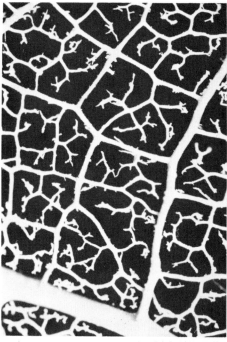

Figure 32   Macerated leaf of *Magnolia sp.* showing part of the closed network of veins.

Leaf Tissues

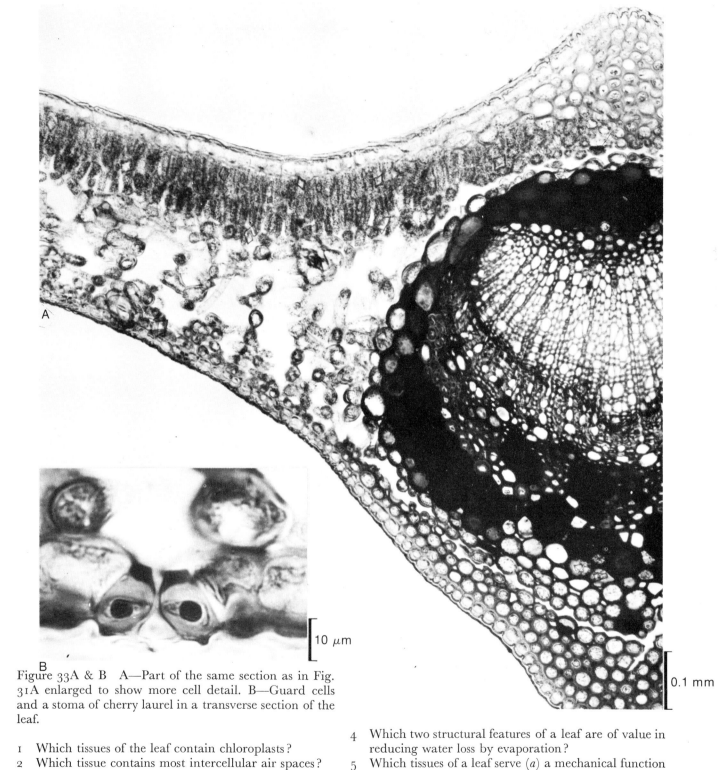

10 μm

0.1 mm

Figure 33A & B   A—Part of the same section as in Fig. 31A enlarged to show more cell detail. B—Guard cells and a stoma of cherry laurel in a transverse section of the leaf.

1   Which tissues of the leaf contain chloroplasts?
2   Which tissue contains most intercellular air spaces?
3   What is the distribution of stomata in cherry laurel?

4   Which two structural features of a leaf are of value in reducing water loss by evaporation?
5   Which tissues of a leaf serve (a) a mechanical function and (b) a conducting function?

Stomata

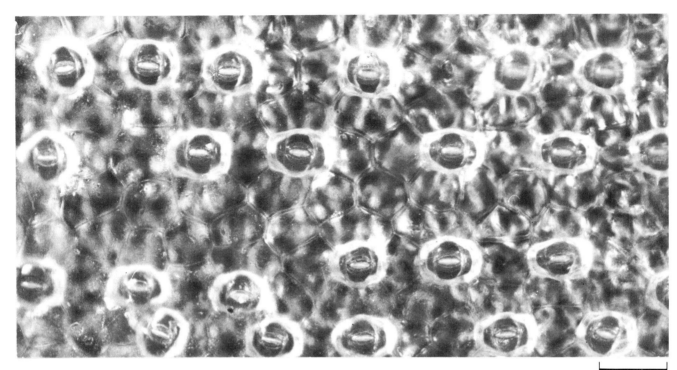

Figure 34   Lower surface of an entire leaf of *Setcreasea purpurea* photographed by reflected light. Notice how the epidermal cells adjacent to the guard cells of a stoma partly cover them.

0.1 mm

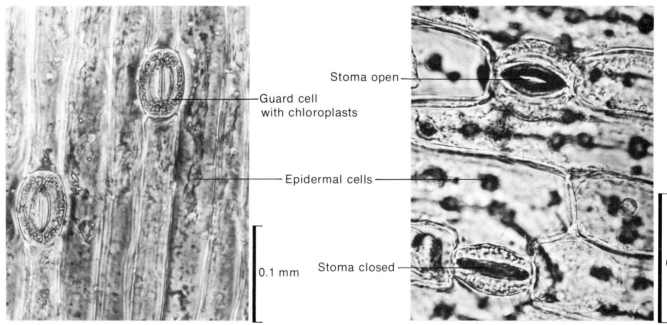

Guard cell with chloroplasts

Epidermal cells

Stoma open

Stoma closed

0.1 mm

0.1

Figure 35   The lower epidermis of bluebell (*Endymion nonscriptus*) cleared of mesophyll and viewed by phase-contrast.

Figure 36   Upper epidermis of red hot poker (*Kniphofia sp.*) also cleared of mesophyll. The black dots in each epidermal cell are hairs.

Stomata and Mesophyll

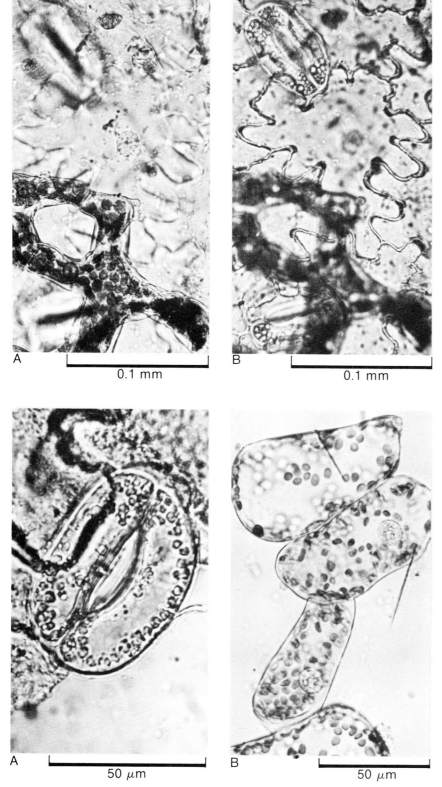

A

0.1 mm

B

0.1 mm

A

50 μm

B

50 μm

Figure 37A & B Partly cleaned lower epidermis of lesser spearwort (*Rananculus flammula*) photographed at two different focusing levels. A blue filter makes the chloroplasts appear black.

1 What is the shape of guard cells in (*a*) surface view and (*b*) in transverse section?

2 From Fig. 34 calculate the number of stomata per cm².

3 In what way do guard cells (*a*) resemble mesophyll cells and (*b*) differ from epidermal cells?

4 Looking at Fig. 36 do you think guard cells open their stoma when they are turgid or flaccid?

5 Can you suggest a reason for their being turgid when illuminated?

6 What is the spatial relationship between stomatal pores and intercellular air spaces of the mesophyll?

7 Two mesophyll cells in Fig. 38B have obvious nuclei and nucleoli; can you find the nucleus of a third cell in this photomicrograph?

8 Which cells on a previous page are most like these?

9 Can you now make an extensive list of the ways in which leaves are adapted to carry out (*a*) photosynthesis and (*b*) transpiration?

Figure 38A & B A—Guard cells surround a stoma of lesser spearwort. B—Mesophyll cells scraped from a leaf of *Kniphofia sp.*

Woody Stem, T.S.

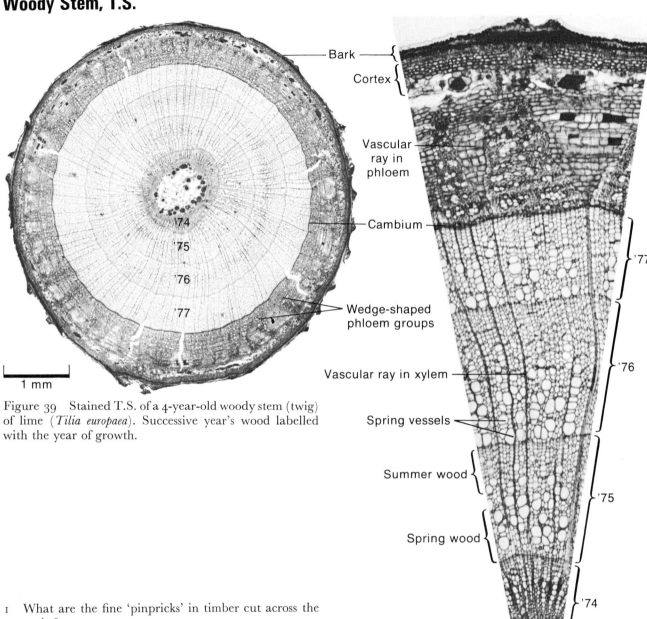

Bark

Cortex

Vascular ray in phloem

Cambium

Wedge-shaped phloem groups

Vascular ray in xylem

Spring vessels

Summer wood

Spring wood

Pith

'74
'75
'76
'77

'77
'76
'75
'74

0.5 mm

Figure 39   Stained T.S. of a 4-year-old woody stem (twig) of lime (*Tilia europaea*). Successive year's wood labelled with the year of growth.

1   What are the fine 'pinpricks' in timber cut across the grain?
2   What tissue in Fig. 40 has divided to form radial rows of cells?
3   Which part of a year's wood has the larger vessels?
4   To what annual event in the lime tree can you relate the development of these vessels?
5   Why do vascular rays appear darker than vessels?
6   Why do vascular rays become wider as they run through the phloem?
7   Why does phloem not accumulate in girth as much as does xylem?
8   What is the value of bark to the plant?
9   Locate vascular rays in Figs. 41 and 42. Why do they appear so different in the two sections?

Figure 40   A part of Fig. 39 more highly magnified

1 mm

Woody Stem, L.S.

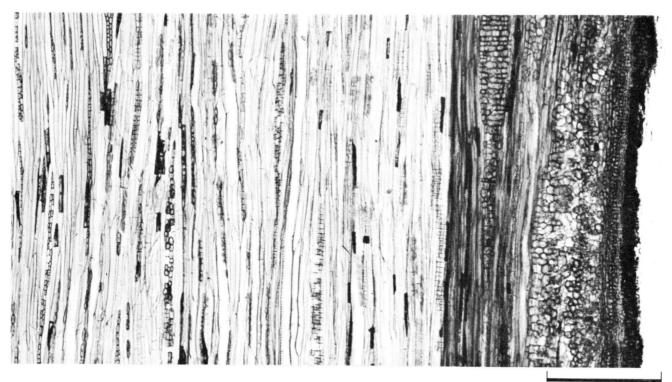

Figure 41   L.S. of a lime twig cut tangentially (T.L.S.).

0.5 mm

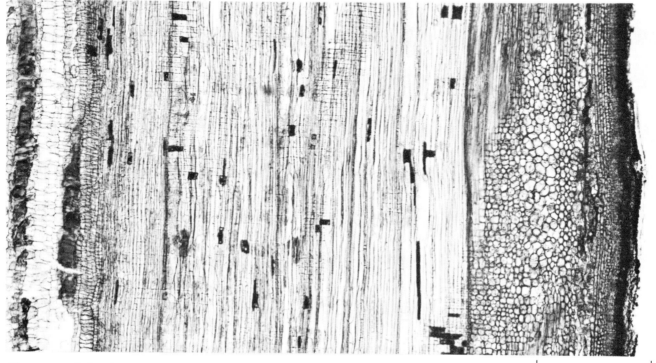

Figure 42   L.S. of a lime twig cut radially (R.L.S.).

0.5 mm

Elder Stem

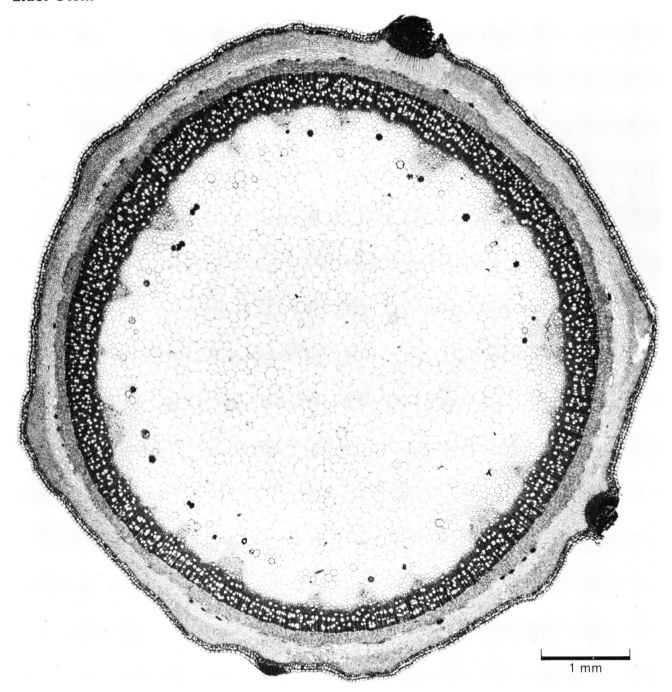

1 mm

Figure 43 Stained T.S. of a woody stem of elder (*Sambucus nigra*).

Examine Figs. 43–45 thoroughly.

1 Can you find in Fig. 43 the extensive pith, xylem, phloem, cortex, epidermis and the three dark external patches known as lenticels?

2 In Fig. 44 can you locate the cambium between xylem and phloem? External to the cambium can you locate another layer of living cells which is producing daughter cells in radial rows? This is the cork cambium.

3 Is the stem's epidermis still intact?

4 What is the dark layer called which is forming under the epidermis?

5 What happens to the cells of this layer in a lenticel?

6 What is the function of (*a*) bark and (*b*) lenticels?

Bark and Lenticels

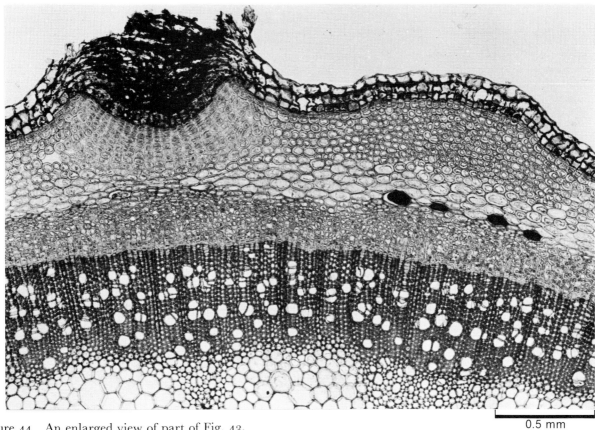

Figure 44   An enlarged view of part of Fig. 43.

0.5 mm

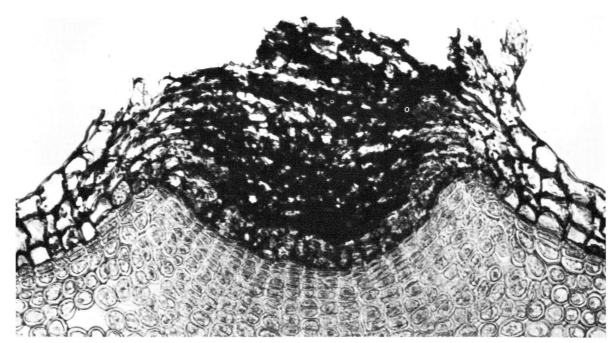

Figure 45   A further enlargement of Fig. 43.

0.1 mm

Specialised Leaves

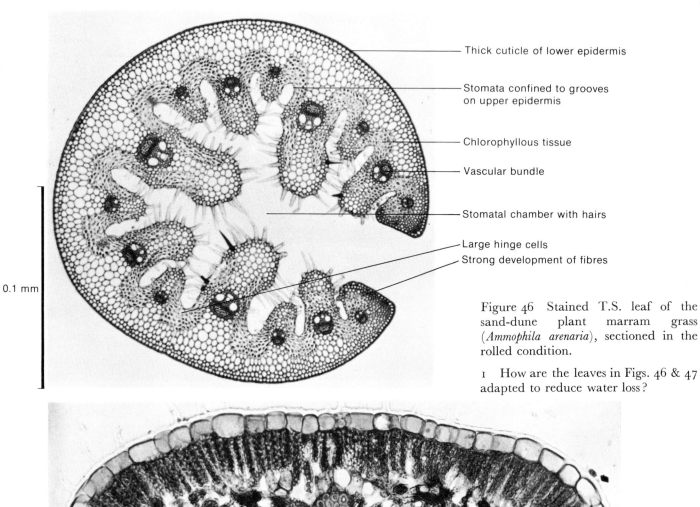

Thick cuticle of lower epidermis

Stomata confined to grooves on upper epidermis

Chlorophyllous tissue

Vascular bundle

Stomatal chamber with hairs

Large hinge cells

Strong development of fibres

0.1 mm

Figure 46 Stained T.S. leaf of the sand-dune plant marram grass (*Ammophila arenaria*), sectioned in the rolled condition.

1 How are the leaves in Figs. 46 & 47 adapted to reduce water loss?

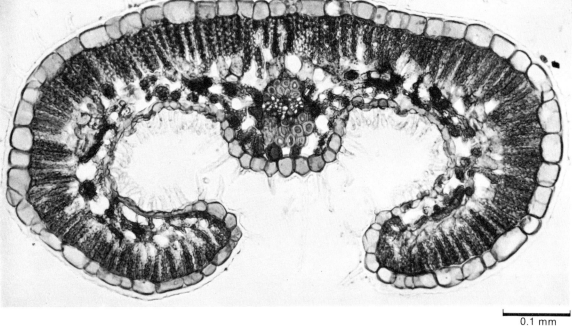

0.1 mm

Figure 47 Stained T.S. of the permanently rolled leaf of *Erica cinerea*, a moorland plant.